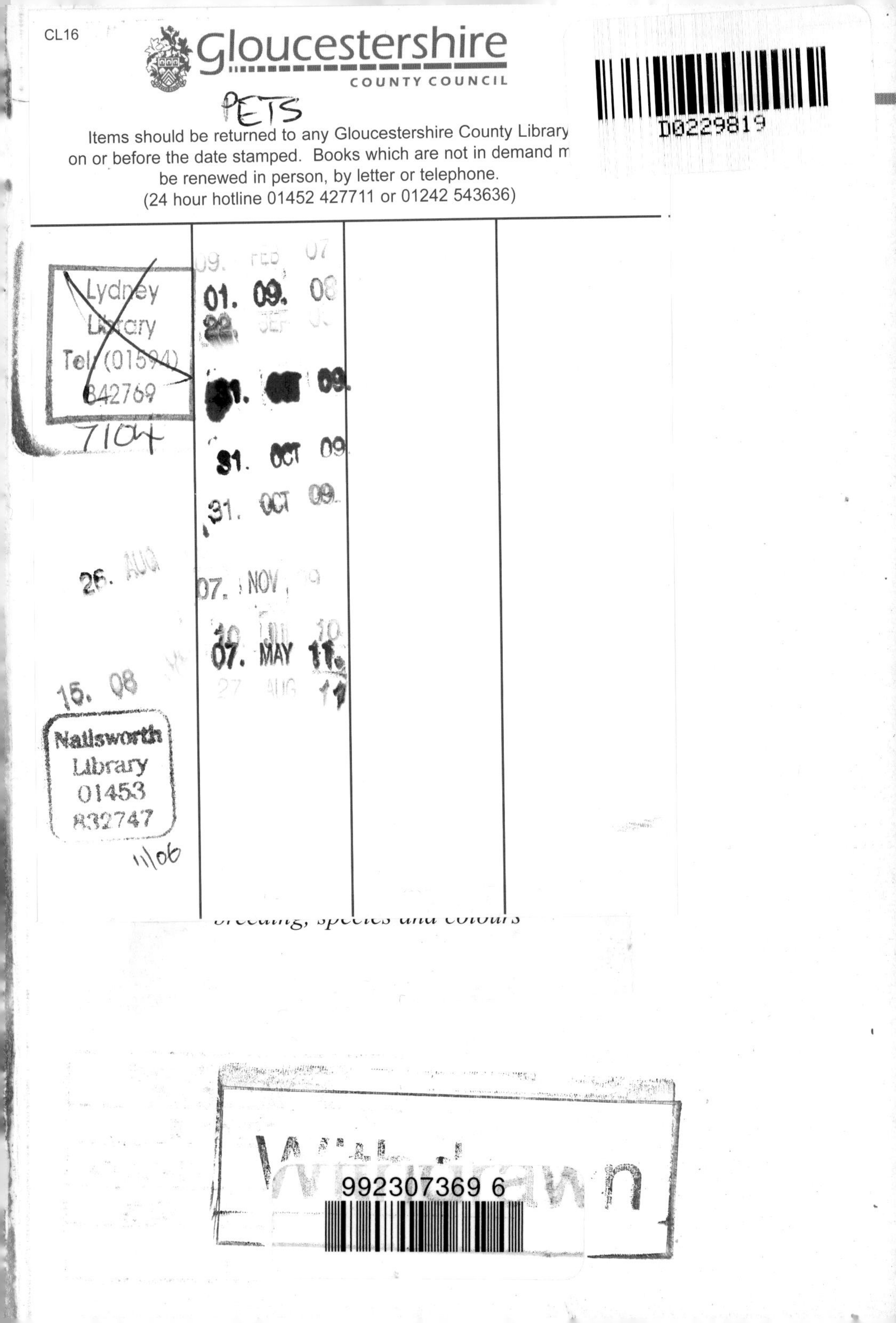

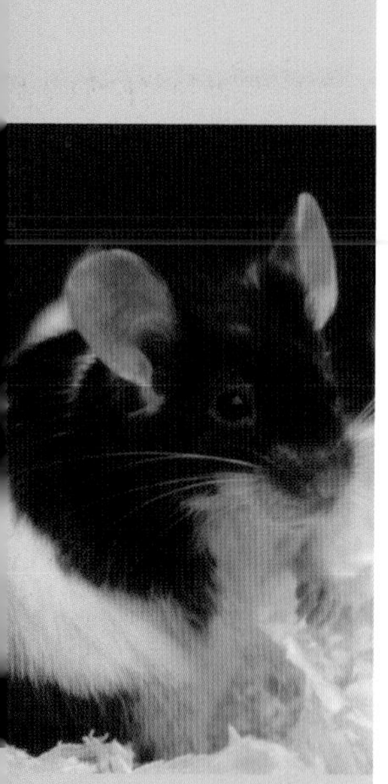

# Contents

# Foreword

Whenever you're planning to buy an animal, it's important to get plenty of information well in advance, even if your preference is for a small animal like a mouse. Is the animal right for your family? How much work is associated with caring for it, and how much will that cost? Can you cuddle it or only view it from a safe distance? This book will give you an overview about the origins of mice, feeding, care, reproduction and the most common ailments and problems with mice. It is also intended to be a guideline for responsible buying and keeping of mice.

A separate chapter is devoted to other species. These are mice closely related to the fancy mouse and kept as pets by experienced mouse-lovers.

About Pets

**A Publication of About Pets.**


co-publisher United Kingdom
Kingdom Books
PO9 5TL, England

ISBN 1852792167
First printing
September 2003

Original title: *de muis*

http://www.aboutpets.info

Photos:
Rob Dekker,
Rob Doolaard, Kingdom Books and
Judith Lissenberg

Illustrations:
Angela Laban

Printed in Italy

# In general

**In biological terms, the mice we keep as pets are identical to their wild counterparts. But the domestication process has resulted in big differences. Wild mice in the cellar or kitchen are unappealing.**

They're unhygienic, shy and nervous, and we have to fight to control them. But nobody needs to panic at the sight of a pretty cage with a number of mice. Mice kept as pets are exceptionally tame, friendly and clean.

The mouse as a pet is the theme of this book. This type of mouse is also known as the 'fancy mouse'. Here and there we refer to the wild mouse (house mouse), because it's interesting to compare the behaviour of the two types and to see what the differences and similarities between the house and the fancy mouse are.

## History

The fancy mouse is a direct descendant of the house mouse. Both varieties can mate together. Man has bred fancy mice for centuries. The Chinese lexicon 'El Yah' in 1100 BC describes fancy mice kept by humans. Mice also play a prominent role in Greek mythology. In Europe, mice have been used for experiments since the middle ages, usually for medical purposes, so it's safe to assume that people have kept the offspring of these laboratory animals more or less as pets for centuries too.

In England, the origin of keeping small animals as a hobby, people first began to breed mice in different colours.
A Mr. Walter Maxey established the National Mouse Club in 1895. The first mice arriving in England were very probably brought in from Japan by Portuguese sailors.

### Differences between the house mouse and the fancy mouse

| House mouse | Fancy mouse |
| --- | --- |
| Wild and nervous | Tame and calm |
| Frightened of people | Not frightened of people |
| Carrier of diseases | Almost never carries diseases |
| Rapid movements | Calm movements |
| Can't be handled | Easy to handle |
| Bites when caught | Never bites |
| Smaller and lighter | Larger and heavier |
| Grey, brown or black | Many different colourings |
| Only active at night | Also active in daytime |
| Harmful to man | Useful for man |

## Origins

The house mouse originally lived on the steppes and semi-desert areas around the Mediterranean, the Middle East and Southeast Asia. Long before our era, mice began to live closer and closer to humans. Over time they became a so-called cultural follower. Most sub-varieties of the wild house mouse finally became totally dependant on man. The house

Field Mouse in the kitchen

mouse loyally followed man on his expeditions of discovery throughout the world, and there is practically nowhere on earth where mice don't live. Its unbelievable capability to adapt is characteristic of the mouse. Classic are the descriptions of mice that live in deep-freeze warehouses at a constant temperature of minus ten degrees, and yet are still able to raise their young.

### Rodents

Mice are rodents. As you can see in the diagram on page 12, they belong to the mammals, just like humans, dogs and horses. Rodents actually form the largest group of mammals; of all the species of mammals in the world, more than half are rodents. The best-known rodents are probably mice and rats, but in fact they come in all shapes and sizes. The largest rodent in the world is the Capybara, or "water pig", which can grow to over one metre long and weigh more than 60 kg. The tiny tot among the rodents is the African dwarf mouse, which is never longer than three centimetres. Between these extremes one finds the squirrel, the guinea pig, the porcupine, marmot, hamster and countless others varieties. Contrary to popular opinion, rabbits and hares are not rodents. They are more closely related to hoofed animals, such as the goat. But rabbits and hares do share one significant characteristic with the rodents; they have continuously growing front teeth without roots. Because rodents are constantly gnawing, they grind their front teeth down. Nature found a solution by having their teeth just continually growing. However rodents do run the risk of so-called "elephants teeth" and you can read more about that in the chapter on "Your mouse's health". When choosing a cage or a hutch, remember that rodents have sharp teeth and can easily chew a hole in wood.

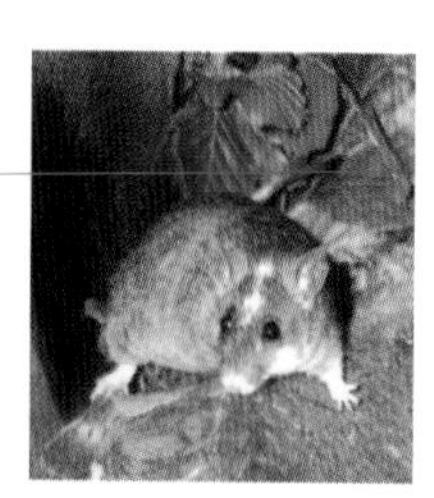

Mouse

### Family tree

A family tree can help to show the position that rodents occupy in the animal hierarchy. All existing animals were first divided into two groups: vertebrates and non-vertebrates. The vertebrates have hard body parts, such as bones and teeth, and are in turn divided in to five classes: fish, amphibians, reptiles, birds and mammals.

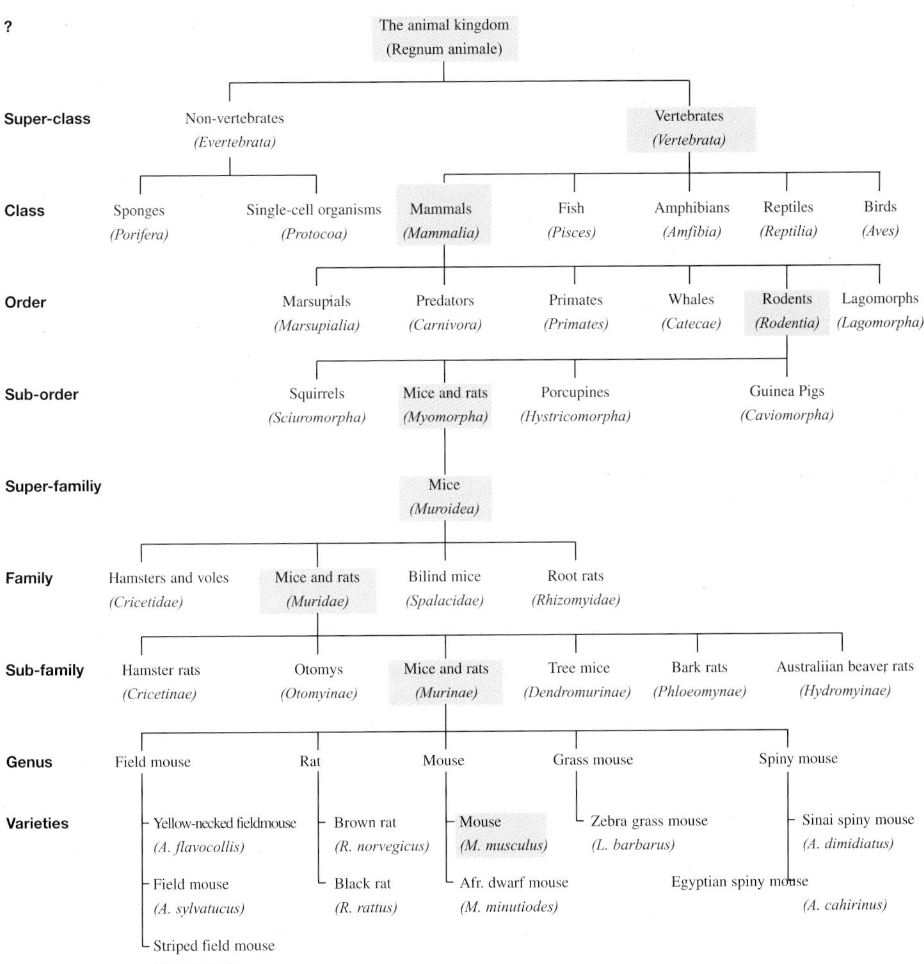
?
The animal kingdom
(Regnum animale)
Super-class
Non-vertebrates
(Evertebrata)
Vertebrates
(Vertebrata)
Class
Sponges
(Porifera)
Single-cell organisms
(Protocoa)
Mammals
(Mammalia)
Fish
(Pisces)
Amphibians
(Amfibia)
Reptiles
(Reptilia)
Birds
(Aves)
Order
Marsupials
(Marsupialia)
Predators
(Carnivora)
Primates
(Primates)
Whales
(Catecae)
Rodents
(Rodentia)
Lagomorphs
(Lagomorpha)
Sub-order
Squirrels
(Sciuromorpha)
Mice and rats
(Myomorpha)
Porcupines
(Hystricomorpha)
Guinea Pigs
(Caviomorpha)
Super-familiy
Mice
(Muroidea)
Family
Hamsters and voles
(Cricetidae)
Mice and rats
(Muridae)
Bilind mice
(Spalacidae)
Root rats
(Rhizomyidae)
Sub-family
Hamster rats
(Cricetinae)
Otomys
(Otomyinae)
Mice and rats
(Murinae)
Tree mice
(Dendromurinae)
Bark rats
(Phloeomynae)
Australiian beaver rats
(Hydromyinae)
Genus
Field mouse
Rat
Mouse
Grass mouse
Spiny mouse
Varieties
Yellow-necked fieldmouse
(A. flavocollis)
Field mouse
(A. sylvatucus)
Striped field mouse
(A. agrarius)
Brown rat
(R. norvegicus)
Black rat
(R. rattus)
Mouse
(M. musculus)
Afr. dwarf mouse
(M. minutiodes)
Zebra grass mouse
(L. barbarus)
Sinai spiny mouse
(A. dimidiatus)
Egyptian spiny mouse
(A. cahirinus)

Fancy Mouse

The latter, of course, were named "the higher species".

The mammals class is again divided into various orders. There is the order of predators, which apart from bears and tigers also includes dogs and cats. The whales fall under a special order. Contrary to what many believe, these animals do not belong in the fish class because their young are born alive. The apes belong to the primate order, while a kangaroo comes under the marsupials. And rodents have their own order, in Latin called Rodentia (wherefore the English word "rodent"). As we already said, rabbits and hares are not classified as rodents, they have their own order, the Lagomorpha.

Field Mouse

The rodent order is then divided into four sub-orders: mice, squirrels, porcupines and guinea pig species. All rats, mice, gerbils and (dwarf) hamsters belong to the sub-order of the mice.

This sub-order of the mice is then again divided into three families: real mice, dormice and jumping mice.

The family of the real mice is split into four genera: hamsters and voles, mice and rats, root rats and blind mice (a kind of small mole). Of course, mice belong to the mice and rats genus (Muridae).

## The difference between mice and rats

Mice and rats are very closely related. In fact, there's only one difference between them: their size. This difference also applies to special varieties of mice and rats. Mice or rat types smaller than thirteen to fifteen centimetres are called mice, anything larger is a rat. The physiological differences between say a house mouse and a brown rat are particularly small. The classification into mice and rats is thus totally artificial and in zoological terms of no significance.

## Good and bad

From the moment that the mouse first met man, it had a huge influence on them. The mouse sought refuge with man, because he provided food and drink. The mouse ate supplies and plants, gnawed holes in clothing and burrowed holes in cellars. So it's no surprise that man tried for centuries to eliminate the mouse. That will probably never be possible. Mice are so clever and cautious and so able to adapt to almost any situation that they're now indestructible.

Apart from the fact that mice do enormous damage to food supplies and other property, there is another huge disadvantage in its long-term attachment to man: the mouse is a carrier of diseases. In the past, diseases such as typhoid and the plague were spread by rats and mice. Because mice are one moment grubbing in refuse and the next moment sleeping in food stores, they can spread disease rapidly.

But mice are vitally important in scientific research. Countless medicines have been developed based on experience with laboratory mice. These white albino mice were brought from Japan to Europe around 1850, and subsequently widely used there for research. The results have been very significant in cancer research, so perhaps one could assert that the harm the mouse has done to man is outweighed by the benefits this little animal has brought with it.

Egyptian spiny mouse

Brown rat

# Buying a mouse

**Fancy mice are nice animals to keep as a pet. Their quiet and affectionate nature makes them suitable as pets for very young children, as long as these are old and sensible enough to be able to handle a mouse carefully.**

But the fancy mouse is by no means just a children's animal. Adults can also derive much pleasure from a group of climbing and playing mice.

## Things to consider in advance

A small and inexpensive rodent like the fancy mouse does involve some costs and effort. It needs a cage, food and care. Caring for one animal can involve (a lot) more time than another animal, but care is something that is needed every day, even when you come home tired from work or school, and during holidays. Before buying a fancy mouse (or any other pet), it's important to discuss this with the whole family. Anyone who plans to buy an animal should get as much information about it in advance. Reading this book is a good step in the right direction!

However sweet mice are, they do have their own needs in terms of care, food and environment. Discuss things with your family and ask yourself the following questions: Is a fancy mouse a suitable pet in your family situation? How intensive is the care it needs and do you have the time for it, over the long term? What does it eat, and what kind of cage does it need? Will you keep it alone or will it feel better in a pair or a group? How much is it going to cost to buy and care for (including vet's bills) and can you afford that? To avoid disappointments later, get the answers to these questions before buying your pet.

In any event don't be tempted by 'love at first sight', because buying on impulse is guaranteed to bring disappointment with your new pet later.
If you're buying mice for a child, make solid agreements in advance about who is going to feed them and clean their cage. Practice shows that children often promise a lot in their enthusiasm, but don't always keep their promises over time. Don't expect too much, especially with very young children: they're certainly not old enough to take responsibility for a pet on their own.

Before taking a mouse home, be sure you've got proper accommodation for it (a cage, container or other accommodation). After all, you can't keep a rodent in a cardboard box.

## What species?

There are many different mice species that can be kept as a pet. These animals all have their own wishes, needs, problems, charm and character traits. Most other species varieties need specialist knowledge. If you have little or no experience in keeping rodents, it's best to start with a normal fancy mouse.

## One or more?

The question as to whether to keep mice alone or together is very simple to answer. In the wild, some animals are social creatures

Egyptian spiny mouse

and live in groups, families or as a pair. Other animals live alone, sometimes completely alone. They seek a mate to breed and then go their own way again.

So it's important to keep animals that live alone in the wild alone in captivity too. Animals that originally lived in groups or as a pair are happier pets when they have a mate. Almost all mice are very definitely herd animals. They will feel very unhappy sitting alone in a cage. Mice may not be able to tell you about their mood in words, but they can certainly do this with their behaviour: a lonely mouse will become listless and apathetic over time. If you're planning to keep mice as pets, you would be well advised to keep two or more animals of the same sex. There's then no risk of being overwhelmed with young, and, fortunately, it's not difficult for mice to get used to each other.

## Where to buy

You have plenty of options where to buy a mouse.

Most mice are sold by pet shops, which in itself is a good thing, as generally pet shop owners know how to look after the animals they sell properly.

However there are always some shops that are not so good. You can often spot what kind of a shop you're dealing with. Are the cages clean? Do all the animals have clean water? Are there too many animals in a small tank? Are they selling animals that are wounded or appear sick?

Most pet shops obtain their rodents from enthusiasts or serious breeders. These animals are healthy and usually used to human hands. A good pet shop owner will also frequently handle his or her young animals.

Unfortunately, as well as serious breeders and animal lovers, we have our share of rogue breeders in this country. These are people who try to get rich quick by breeding as many mice, hamsters, rabbits or other pets as possible, often keeping the animals in disgusting conditions. They never, or hardly ever, take care of hygiene or animal welfare, and in-breeding is the order of the day. One of the major disadvantages of these "breeding factories" is that the young are separated from their mother far too early, because time is money after all. The young are nowhere near strong enough, and sooner or later become seriously ill. Never buy a mouse (or other pet) that is still too young or too small.

However hard it may sound, never give in to the temptation to buy such an animal hoping to give it a better home. You're really not doing any good. The more animals these dealers can sell this way, the more they will keep "in stock". They don't care why you'-re buying the animal, just as long as you buy it. But if nobody buys their tiny sick animals then they can't make any profit either, and they then have to decide either to stop trading or to start taking better care of their animals.

### Things to watch out for

If you're planning to buy a mouse, watch out for the following points:

- The animal must be healthy. A healthy mouse has bright eyes and is lively. Sexual organs must be clean and the animal must not show signs of wounds or scars. Its coat must be smooth, clean and glossy. Look out for any lumps or swelling.

- The mouse must not be too young or too small. During the first weeks of its life the young animal gets resistance to disease through its mother's milk that is vital to its health.

- The animal must also not be too old. Mice have a relatively short life. Buying an older animal may mean you will only be able to enjoy it for a short period. Also, grown mice are more difficult to tame. You can recognise an older animal by its yellowish belly.

- Check whether your mouse is really the same sex as the shop

assistant tells you. Despite the fact that the difference is very easy to see, mistakes are often still made on this point: two "females" often suddenly produce babies later.

- Make sure the animal is not too thin or too fat. A fat female is almost certainly pregnant!

Don't buy a female from a mixed group. There's a very good chance she is pregnant

### Selling young mice

However much you like breeding mice, sooner or later you can't keep them all. Then you must sell, swap or give away the young.

You can sell young (and older) animals to a pet shop. In this case you never know where they will end up. In any event, look for a serious pet shop that provides broad and correct information when selling animals.

You can also sell the young to private individuals. You might notify a mouse owners' club (see the chapter on Addresses), place an advertisement in the newspaper or, preferably, sell or give them to acquaintances. Always give the new owner(s) good information about feeding and caring for mice.
Whatever you do, never set animals free into the wild!

### Catching and handling

Fancy mice are almost always quiet and can easily be picked up out of the cage by hand. Other mouse varieties, such as grass mice and the African Dormouse can be very nervous and extremely fast. Animals that are not used to being picked up by human hands are always frightened by the experience. When animals are very wild you should use a jam jar to catch them. It's important to catch the animal as fast as possible. Other species, particularly, can be very sensitive to stress. A long chase can tire them out and cause them to go into shock. An animal in shock will lay flat on its chest with paws outstretched and make epileptic (jerking) movements.

### Transport

When you buy, or are given, a mouse, you have to get the animal home. In many cases this is done in a cardboard box. This is not the best solution. It would not be the first time (and won't be the last time either) that a mouse gnaws a hole in such a box and goes off on a journey of discovery in the shopping bag or the car. So it's better to get a transport container in advance, and you can buy one at any pet shop. These containers are too small to be permanent housing, but very suitable for the first journey home or a trip to the vet's. You can also use it to house your animal temporarily when you're cleaning its cage.

# A home for your mouse

**To be able to keep a mouse responsibly and give it a comfortable home, it's important to take a look at how they live in the wild.**

Even if your mouse lives in a hutch or a cage at home, it is still possible to get close to their natural living conditions, making the animal feel as comfortable as possible.

### In the wild

In the wild, mice live in widely differing conditions. They are found in (sub)tropical areas, but as we already discussed, colonies of mice have been found in deep-freeze stores at or below freezing. A mouse is an animal that can adapt to just about any situation, but despite this, there are patterns to be found in the life of a mouse: house mice climb and clamber a lot, love to creep through small holes and into tiny cavities and live in a family.

### Housing in captivity

Looking at the life of a mouse in the wild, we can reach the following conclusions: a mouse likes live together with other mice, so give it company but be careful you don't get unwanted offspring. Mice love to climb and clamber, so a bare home with only floor litter is not for them. Give them a home with plenty of opportunities to climb and play. Taking this into account, they need a pretty, well looked-after and safe mouse home that should at least fulfil the following conditions:

- their home should keep its resident(s) inside. That means a properly closing door, and no chinks, small holes or wide spaces between bars. And remember that a mouse has razor-sharp teeth that will devour anything that is not made of glass, stone or steel.

- their home should be safe for animals and humans. No spikes or other objects, which may harm them, sharp glass edges or lids that fall inward.
- it should be easy to clean, there should be no corner or holes that prevent you from cleaning parts of their home.
- It should be made of a material that won't absorb moisture or smells. Wood is unsuitable for making an animal cage unless it's been treated with a water-repellant coating. When animal urine soaks into wood, it will start to rot and smell. Glass or plastic, on the other hand, is ideal.
- the opening must be wide enough for you to be able to access the entire area of their home. Not just for cleaning purposes, but also to be able to catch the animal if necessary. If the door is too small the animals may be able to get into a corner that you can't get at.
- However small your animals are, their home must be well ventilated. When cage litter is soaked in urine the ammoniac smell can hang around in the bottom of the container. With insufficient ventilation, this can be harmful to the animals.
- There should be areas that the animals can withdraw to in peace.

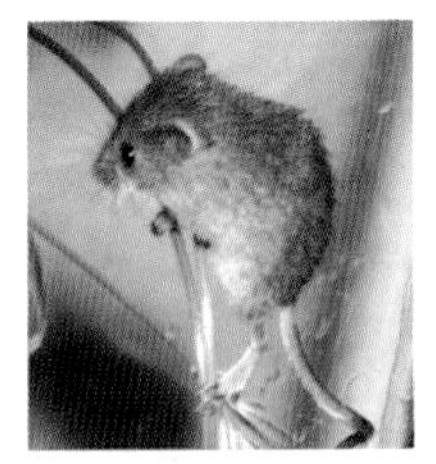

Harvest Mouse

## Types of cage

You can keep your mice in various types of cage. All have their advantages and disadvantages. Let's look at them one by one:

### Wire cage

The majority of cages sold in pet shops are wire cages. They usually consist of a plastic base on which a cage of metal wire sits.

The big advantage of this type of cage is their good ventilation: fresh air can get to the animals from all directions. However, a disadvantage is that draughts can do the same.

Some wire cages have a very small opening. If you need a larger opening, then you have to lift the whole top part off the base and your mice can easily escape in all directions. The better ones have a small opening for feeding, but you can also open the top leaving the walls standing.
Another advantage is that this type of cage is very light and easy to clean.

Because fancy mice don't usually burrow, a low base is not such a problem as with fanatical diggers like gerbils. But you should consider that with a wire cage with a low base, litter might fall out around the cage.

**Plastic or glass containers**
Rodents are frequently kept in old aquariums or plastic containers with a gauze lid. In such a closed container the animals aren't bothered by draughts, but ventilation is not optimal either. So you need to clean the cage litter frequently, otherwise the animals live in ammoniac fumes. A lid of glass or plastic sheet is absolutely out of the question because it allows no ventilation at all.

Plastic containers have the disadvantage that they quickly become unsightly, because they are easily scratched. The toilet corner in these containers can also become corroded and rough, making it difficult to keep clean and bad for hygiene.

Glass containers are available in various forms. One-piece containers are easily kept clean, but are also heavy, and if one crack appears you might as well throw it away. There are also aquarium types with a metal frame holding panels of glass. In the past these panels were fixed with putty, which would, of course, not dry out when the aquarium was filled with water. However, old putty tends to crumble in a dry mouse container. The animals can gnaw at it and the panels can become loose. So containers with putty are unsuitable. Nowadays, glass panels can be fixed with silicone, which is an easy job for any do-it-yourself enthusiast. You can also make a glass cage without a frame.

The glass panels are glued together with silicone and after one day the container is so strong that you could actually fill it with water. This type of cage can easily and cheaply be made at home. The silicone bead should not be too thick, otherwise the animals will gnaw at it, and the corners should be polished smooth or protected with plastic corner strips, because they can easily cause injury.

**Rodent paradises**

Pet shops sell pretty, elaborate rodent paradises. Some have a number of wire houses on top of and beside each other, others have a complete burrow system in plastic. To children these appear to be an exciting place for a mouse and are naturally very attractive. But they're not ideal. The burrows and caves are difficult to clean and poorly ventilated.

**Laboratory pen**

Some people keep their mice in laboratory pen. These are low plastic containers with a metal grille as lid. A laboratory pen is ideal for housing a lot of animals with the minimum of work, which is the intention in a laboratory, but you have very little contact with the animals. It is also questionable whether they feel happy in such a boring home.

## Cage litter

For years wood shavings have been used in animal cages. This is often called sawdust but is actually shavings.
Sawdust absorbs moisture exceptionally well and hardly smells, but a disadvantage is that it also usually contains a lot of dust. Investigations in recent years have shown that this dust can seriously bother rodents. There are now many other types of cage litter on the market that are "healthier" for animals.

**Sawdust**

As we have said, sawdust is not especially suitable as cage litter. Most animals (including mice) get this dust in their lungs and, over time, it can cause infections. Now that the dust problem is generally recognised, some types of sawdust are cleaned better by the manufacturer. But for mice, another type of cage litter is preferable.

**Hay**

Rodents like to use hay as nesting material and to chew on, however it does not absorb moisture well and is thus not really suitable as cage litter.

**Straw**

Straw is much too coarse to be suitable as cage litter or nest material for rodents. There is a product on the market which is made of shredded straw. Russell Rabbit cage litter is wonderfully soft and ideal as nesting material. It is less suitable as cage litter, as it absorbs little moisture.

**Cat litter**
There are probably a hundred different sorts of cat litter on the market. Some are suitable to keep rodents on, especially those made from maize. These absorb plenty of moisture and can do good service. Cat litter made of stone or clay is less suitable, mainly because it can become dusty.

**Pressed pellets**
In recent years various cage litters have appeared on the market that consist of pressed pellets. Some types have sharp edges and don't seem very comfortable.

**Sand**
Some rodents like to live on sand. A disadvantage as cage litter is that sand doesn't store warmth. So, on it's own, sand is not suitable as a cage litter for mice.

**Shredded paper**
There are also various types of shredded paper on offer as cage litter. These shreds are ideal to play with and can be used as nest material. But they absorb much too little moisture to be used as cage litter.

In conclusion, use a cage litter that easily absorbs moisture, in combination with a soft, insulating nesting material.

## The interior

You can also fit out a mouse's home with various articles from the pet shop. There are literally hundreds of different mouse toys in pet shops. Let's look at the advantages and disadvantages of a number of popular rodent articles:

**Mouse huts**
There are countless sorts of mouse huts on sale, made of plastic or wood. Because mice gnaw but not as much as gerbils or hamsters, even plastic huts may last a while. Mice like to have one or more huts. You can even make a hut from a flowerpot standing upside down. You just have to knock a semi-circle out of the edge to make a little doorway.

**Wheels**
Opinions vary on the usefulness of a wheel. Some people insist that these provide plenty of recreation for a mouse. That's surely true, but on the other hand a wheel does force the animal into monotonous activity that may result in psychological disturbances. The fact is that mice use their wheels a lot, but they can also cause accidents. There are no brakes and a mouse can easily get stuck between the wheel and its uprights.

**Straw articles**
Mouse huts, tunnels and balls made of straw have appeared on the market in recent years. These products are made of woven straw and hay held together by wire. These are ideal toys for mice! They can climb and tunnel in and

around them and gnaw at the straw and hay. After a while the straw hut or tunnel is finished with and you just need to take the wire skeleton out of the cage.

### Climbing and clambering

Because mice love to climb and clamber, it's important to divide their home into several 'floors'. This gives them much more living space. The space between floors does not need to be that high. Fancy mice almost never jump. You can connect the floors together with steps, ladders or climbing rope. Your mice may also climb up the bars of their cage to get a higher floor.

If you're not keen on ready-made plastic objects, you can fit out your mice's home really nicely with rocks and twigs, giving it a pretty and natural look.

### The best place

Take care when picking the place to stand your mouse cage. Places where big temperature differences can occur, such as near an oven or radiator are not suitable, nor is a window sill which is sometimes in the full sun. Mice like to party, but living permanently on top of a loudspeaker is too much of a good thing. The garage or garden shed is also not ideal; after all you want to see your animal from time to time! Apart from that, these places are too quiet, too dark and often too damp. The best place for the cage is in the living room or a child's bedroom, out of the sun, away from draughts and, where possible, off the ground on a (low) cupboard or table

### House-training

In principle, most rodents are house-trained by nature. They don't like to foul their own nest and will always do their business in the same corner of their home. This can be very practical, as in some cases you don't have to clean out the whole cage, but just have to clean out the toilet area.

### Scents

It's a fact of life that animals often smell and sometimes (particularly) unpleasantly. The so-called drier (special) mice hardly smell at all, but the scent of fancy mice can indeed become unpleasant over time. You can keep this under control by cleaning its home regularly. Experience has shown that mice will smell much less if you only replace their cage litter, rather than cleaning  the floor of the cage itself. Then they don't need to deposit their scent, as the cage already carries it.

# Feeding your mouse

**Day-in, day-out for years on end, rodents are fed the same thing: mixed rodent food. However, research into the feeding habits of rodents in the wild has shown that they generally need a quite different and more varied diet.**

Mixtures

### Feeding in the wild

In the wild (insofar as this still applies to house mice) mice will eat practically anything. They will eat whatever is available at the time and place. This may be dried meat, but also grain, vegetables or fruit. When they're hungry they will even eat cloth. In general, they enjoy an extremely varied diet.

### Feeding in captivity

When you consider their feeding habits in the wild, it's illogical to feed mice only a mixture of oats, grain, barley and grass pellets all the time. These "old-fashioned" foods contain almost no animal content and thus hardly any proteins. So a lot has changed in the field of rodent foods over the last few years. Major manufacturers have developed special foods for almost every species of small rodent, including the fancy mouse. Check to see that there are not too many green pellets in the food. Manufacturers stuff their vitamins and minerals into these little green sticks, but mice generally won't eat them.

Whenever you buy any food other than a special mouse food, take care that its structure is not too coarse. Rabbit food is certainly not suitable, nor are grass pellets (in the form of little dark green sticks). If you use a normal mixed rodent food, you can supplement it with dry cat or dog foods, and possibly weed or grass seeds from the bird-food shelf.
When buying food, look out for the date of manufacture. The vitamins in the food stay effective for about

three months, after which they quickly lose their goodness, so never buy too much food at once.

Just like humans, animals also like variety, but never give rodents sweets, crisps, biscuits or sugar lumps. The salt and sugar these foods contain can make them seriously ill. If you want to give your mice a treat once in a while, there are plenty of healthy pet snacks that you can use to put something special on the menu.

## Pressed pellets

Pet shops also carry ready-made foods in the form of pressed pellets. These pellets all look the same and have the same ingredients. Many breeders give their animals such foods, because then they're sure that each animal gets all the nutrition it needs. Apart from that, a lot less food is wasted as the animals don't pick out what they like best, leaving the rest. But the question is whether they really like getting the same food every day.

Pressed pellets

### Vegetables

Mice enjoy most sorts of vegetables, but they shouldn't consume too much. They can cause diarrhoea. But now and again they can be treated to a small piece of chicory, endive, carrot, cauliflower, paprika, cucumber or broccoli, however lettuce and cabbage is best left out of their diets.

### Fruit

Most rodents, including mice, enjoy fruit. What sort is their favourite depends on the animal. Almost every rodent loves apple, pear, peach, raspberries, melon, berries and banana. Most animals find citrus fruits such as oranges and mandarines too sour, but some seem to enjoy them. Just like vegetables, too much of a good thing can be harmful. Take care to remove from the cage any pieces of fruit or vegetables that have not been eaten at once. There is a chance they will start to rot, which could make your mouse ill.

### Eating their droppings

Almost all rodents eat their own droppings from time to time. This is not only normal, but necessary. During the digestion process, vitamin B12 is produced in the intestines. By eating their droppings the animals take in this important vitamin.

Young mice eat their parents' droppings, because they contain the bacteria they need to be able to create vitamin B12 in their own intestines during digestion.

Rodent snacks

## Vitamins and minerals

Vitamins and minerals are elements every body needs to stay healthy. As long as a mouse enjoys a good, varied diet it does not need additional vitamins and minerals. These are in their food. In the chapter Your mouse's health there's more information about the consequences of an unbalanced diet.

Some breeders hang a so-called "mineral lick" in the cage. The animal takes in minerals by licking the stone and it seems that pregnant females, particularly, use them. Sometimes small blocks of limestone are placed in the cage. By gnawing on these, mice get extra calcium and at the same time keep their teeth sharp.

## Water

Mice drink a relatively large amount of water for their size. So it's important for a mouse to always have fresh water available. It's best to give water in a drinking bottle hung onto the outside of the cage with it's spout pointing inside. You can buy these bottles at any pet shop. A dish of water is quickly overturned or filled with sawdust or other dirt.

Water bottle

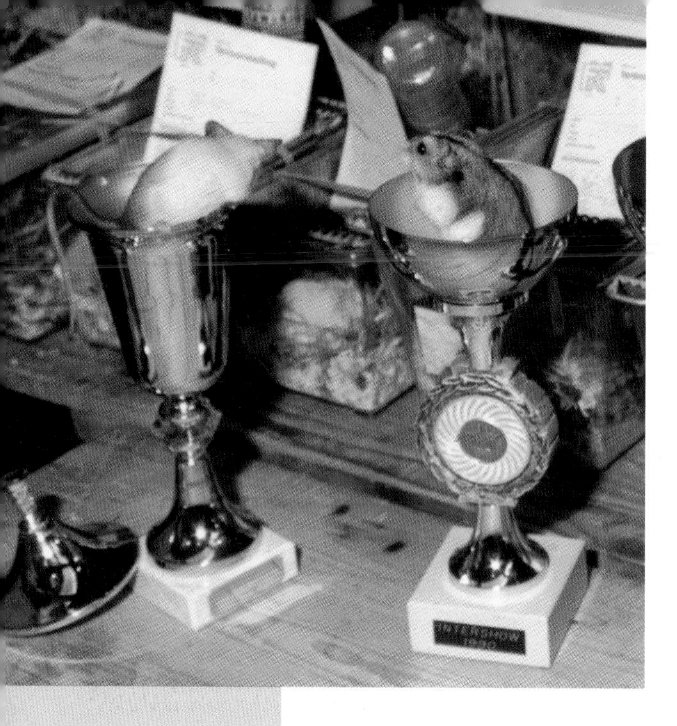


# Shows

**Many people breed fancy mice as a hobby. Mice-lovers have a number of containers with mice at home and try to breed beautiful examples from them with which they can win prizes.**

During one-day shows, or even shows lasting several days, judges assess all the animals that have been entered for size, colour, shape and condition. The perfect fancy mouse must conform to the requirements set out in the Standard. This standard describes precisely how each variety of fancy mouse should ideally look. If you're interested in taking up this hobby, you can contact the .........(see the chapter 'Addresses')

Even if you're not planning to breed yourself but are interested in rodents, it's well worth the effort to visit a small animals show. You can get a lot of information and the breeders present often have good animals for sale.

## The Standard

The standard describes how rabbits, fancy mice and other small rodents such as the Guinea Pig, the Dwarf Hamster, the Mongolian Gerbil and the fancy rat should ideally look. An animal entered for a competition can earn points in seven categories. In the table you can see how many points a fancy mouse can score in each category. Points are deducted for any defects depending on their seriousness. The animal that finally scores the most points is the winner and earns the title 'best in show'.

Fancy mice come in many more colourings and markings than those included in the standard. But a colouring or marking is only officially recognised if it's in the

standard. For example: black and blue mice are described in the standard. If a mouse is entered that is neither black nor blue, but something in-between (dark blue, for instance) , then it doesn't meet the requirements and will get a medium or poor score in the 'colour' category.
One can view the animal as a 'poor' black or blue mouse. It is also possible to enter the animal as a 'new colour' for the standard, but not every mixture of two colours will automatically be approved. There are a few requirements, which a colour or marking must meet to be included. At least four animals with the new colour or marking must be entered for the federation's show. There they are judged and possibly approved by the standard committee. The first step is preliminary recognition. If, after three years, there are enough animals with the new colour or marking, it is then fully recognised.

## Judging

A number of different types of competitions for mouse-lovers take place. Most take place during shows that are held across the country. These are sometimes small and short meetings, larger shows can last several days. In some cases judging is open to the public. At larger shows it usually follows the opening ceremony.
In general the judging sessions can be viewed freely, Many small animal clubs organise their own club competitions, which are usually also open to non-members.
During a competition, the judge sits behind a small table, on which there's a plank covered in carpet. The animal to be judged is set on the plank. You can recognise the judge in a competition by his or her white jacket. You don't become a judge just like that, several years of study are needed before an experienced breeder may call himself a judge.
A judge will assess each animal in the following categories:

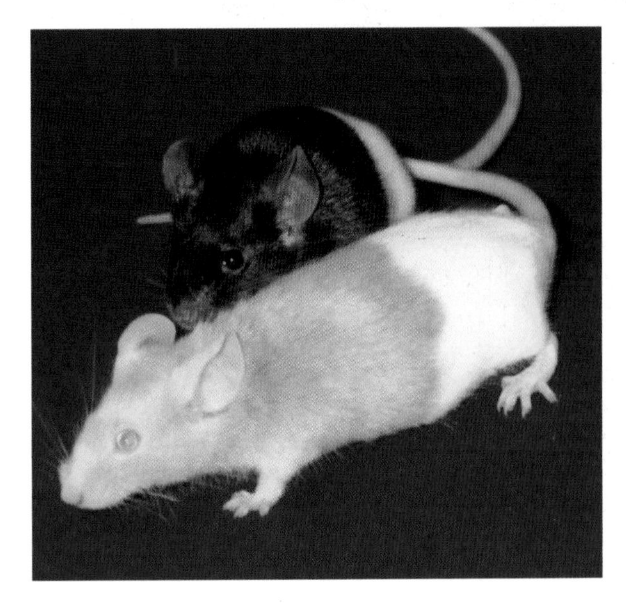

Rumb white

### Type and build

This part of the standard describes a mouse's build. A quote: "A mouse should be long and slender, with a wide breast. The belly should be dry and somewhat

drawn in. The tail may not be thin and should be approximately of the same length as the body. The tail should flow smoothly from the body. Defects for which an animal may lose points are: deviations in type and build, tail too short or too thin, lumps or bends in the tail, missing claws and wrinkles on the belly"

**Size**

The primary factor in this section is the category the entered animal is assigned to: young or old. For the purposes of a competition, a mouse is regarded as adult from an age of ten weeks. In the size category the following applies: the bigger the mouse, the better. But it must stay slender and finely built. An adult mouse should ideally weigh 45 to 60 grams. The weight can vary per breed.

**Coat and hair condition**

Mice may possess various hair structures, but not all are officially recognised. You can read more about that further in this chapter.

**Head, eyes and ears**

The head should begin with a wide skull and run conical to the nose, but it should not be too pointed. The large ears stand upright and are open, with a wide base and well rounded. The large ball-shaped eyes stand out a little. Defects: Pointed head, small ears, folds in or damage to ears, and sunken eyes.

Broken marked

**Topcoat and belly colour**

The topcoat is understood to be the surface colour on the animal's back. The word 'surface' is very important here. A hair usually consists of different colours. The base of a hair (at the body) will have a certain colour. In certain cases the intermediate part of the hair will have its own colour. The top and belly colour are important because the colour of the hair tip is obviously the most visible. The various top and belly colours are described in the colouring descriptions.

**Middle and base colour**

After reading the above, it should be clear what is meant by intermediate and base colour. These colours can be seen by blowing into the coat. The hairs lay back, forming a kind of rosette that clearly shows the base, intermediate and topcoat colours.

**Body condition and care**
This part is an assessment of the general impression the mouse makes. It should be lively and view its world with clear eyes. Sick or wounded animals are regarded as seriously defective, as are visibly pregnant animals.

It should be made clear here that mice sold in pet shops are not pedigree mice. After all, they've been bred as pets, and not for showing. You could call them "medley mice", but they're often the nicest and best-natured pets. But it's worth finding out how close your pet mice come to meeting the standard. Some parts, such as body condition and care, also apply to a pet. Once in a while, look at your mice as if through the eyes of a judge: are they not too thin, or too fat? Is their coat well looked-after?

A newer marking: Brindle

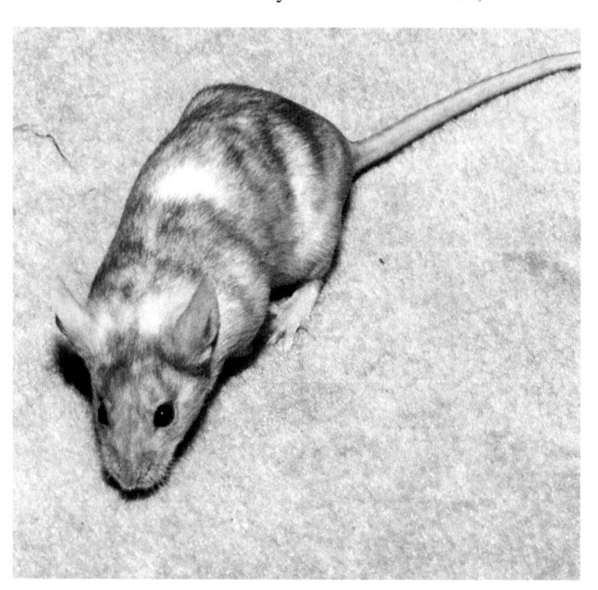

## Hair structures

As we said earlier, mice can possess various hair structures. The recognised structures are normal, satin, longhaired and bristly.
At the moment there are again attempts being made to breed mice with a Rex coat, but this structure is no longer recognised. The Rex coat is a mutation that was produced in the United States in 1945 and was recognised in Europe until 1978.

**Normal hair**
Most mice have a normal coat structure, which is called normal hair. This is a short, tight fur with a slight gloss. Dull, open fur or bald patches are regarded as defects, as is moulting.

**Satin coat**
A layman can't always recognise a satin coat. This type of coat is of hairs that are almost transparent lending the coat a kind of satin gloss. This causes a splendid effect with certain colourings. Gloss, density and structure are very important for satin coated mice. Extra points can be won for hair structure and gloss. Points will be deducted for an open coat and too little gloss, as well as for bald or thin patches.

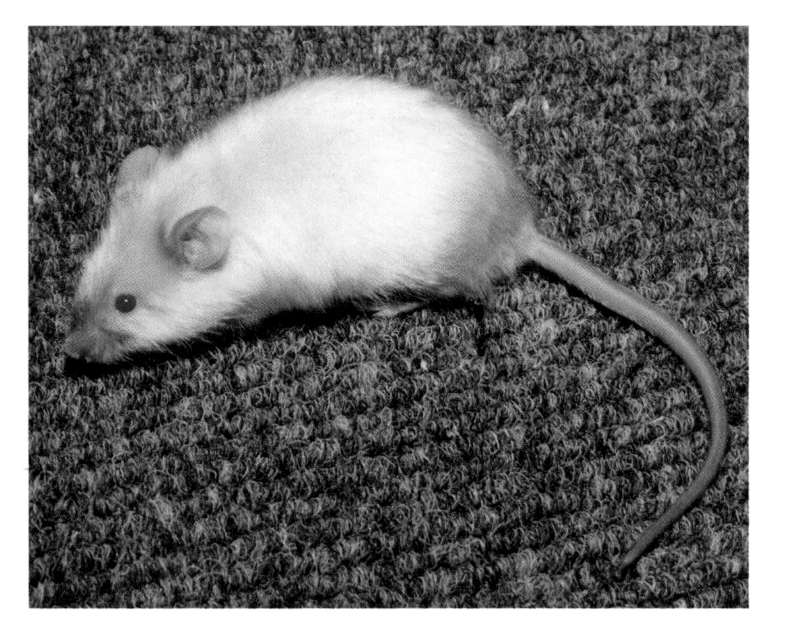

Siamese

**Longhaired**
The rule 'the longer the better' applies to this coat type. Besides the length, the coat's density and structure are key to the assessment. Male mice have the longest hair, which falls to both sides with a pretty parting on the back. There are also long-haired mice with satin coats. Of course, hair that is too short is regarded as a defect, and if it's shorter than one and a half centimetre, then this is a serious defect. Bald or thin spots and moulting are also not permitted.

**Bristly hair**
A very recently recognised hair structure is the bristly hair. After preliminary recognition in March 1992, this is now fully recognised. In the beginning, mice with this hair structure were on the small side, but breeders have managed to largely correct this. The bristle hair is tangled and somewhat longer than normal. A crest runs from the centre of the back to the hindquarters. There must be at least one rosette on each hip, more rosettes win extra points. Besides the normal defects the crest and rosettes are checked for irregularities. They must not stand open or be missing.

## Recognised colourings

Fancy mice come in many different colours, not all of which are recognised. It is generally difficult to describe the exact colouring. That would only be possible with a good colour photo taken and viewed in just the right light.

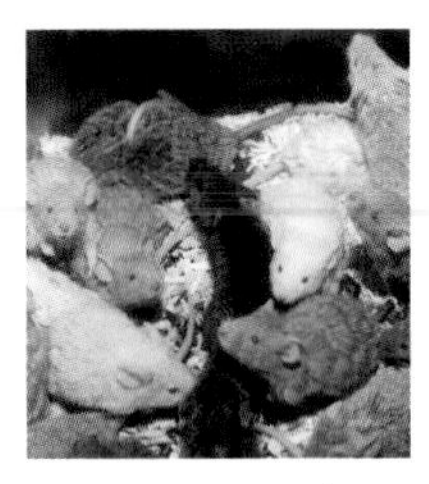
Orange

Despite the fact that judges are well trained and usually experienced, even they don't always agree on the colour of a particular example. Unfortunately, within the scope of this small book, it is not possible to show all colourings with a photo. The following descriptions will describe the colour description in the standard as precisely as possible

## Agouti's

An agouti is a mouse with a blended coat. This effect arises through ticking. Ticking means that brown (or other coloured) hairs carry a black spot. Many animals in the wild have such a blended coat, including the rabbit and several varieties of mice. Why they're called Agouti is not clear: the real Agouti (a rodent in the Guinea Pig family) almost never has ticking.

The Gold Agouti is the colour that comes closest to the colour of a wild mouse (house mouse). The warm golden-brown top colour shows a regular black ticking. The paws are lighter, but must show the ticking. The belly is the same colour as the back, but with less ticking. The eyes are black, as are the whiskers. Tail, ears and soles of the feet are somewhat darker. The undercolour is dark slate blue.

The Cinnamon Agouti is a colour where the black in the hair is replaced by a rich brown, the

colour of cinnamon. The golden-brown topcoat colour is somewhat lighter than on the Gold Agouti. The undercolour is also somewhat lighter.

The Silver Agouti is silver-grey with black ticking. The whiskers are black. Tail, ears and nails are dark. As with many colourings the belly is duller in colour with less ticking. The undercolour is dark slate-blue.

The coat of Argente mice is a mixture of pale yellow and silver blue hairs. The belly is pale yellow, eyes pink-red. Soles of the feet are lightly pigmented, as are the nails.

The colouring Chinchilla is named after the Chinchilla, a rodent of the Guinea Pig family, that can also be kept as a pet and is famous for its pretty fur. On mice the chinchilla colouring is pearl-grey with a dark blue ticking. Ears and tale are somewhat darker than the body coat. Eyes are black, the belly and the insides of its feet are white. The intermediate colour is pearl-grey and the undercolour slate blue.

### Single-coloured mice (Selfs)

Single-coloured mice are called 'selfs' and in contrast to agoutis have a single colour over their whole body. The intermediate and undercolour are as close as possible to the topcoat colour. In general the belly colour is somewhat duller and lighter than the topcoat. Watch out for white hairs in the topcoat of a self; a few white hairs are regarded as a small defect, a lot are a serious fault.

A Black self mouse must be an even deep black over its entire body. Only the belly fur may be somewhat duller in colour.

The colour Blue on many animals is just a diluted black. Just like the black mouse, the blue self mouse is the same colour all-over but a slate blue. Its eyes are dark blue.

Chocolate self mice have the colour of dark chocolate. Eyes, ears, whiskers and tail are dark brown.

The Red self mouse is a warm red all-over. Its eyes are dark brown. Ears and tail are somewhat darker.

An Orange self is totally in a warm orange. The colour may not tend to yellow. Its eyes are red. Like the other selfs, this colour must show no signs of ticking. The ears must contain a little pigment.

Yellow self mice must be all in a warm yellow. The colour must not tend to orange. Eyes are red.

The Dove-grey self mouse is actually black but, due to the effects of so-called red-eye dilution, has become a pleasant grey. This grey must not tend to blue or show a reddish glow. The eyes are red.

The colour Lilac is also a dilution. This is in fact a chocolate mouse with a blue factor. This results in a bluish mouse with a strong reddish glow. Its eyes are blue with a red glow caused by the brown factor.

Champagne self mice are bred from chocolate mice with a red-eye dilution. Over its light beige topcoat lies a reddish glow. The eyes are red. A pallid colour is regarded as a small defect, the lack of a reddish glow is a serious defect.

A Crème self mouse is actually a red mouse into which the chinchilla factor has been bred. The soft creamy topcoat must be as even as possible. The eyes are black. A different eye colour (which sometimes occurs) is counted as a serious defect.

White red eye

There are two varieties of white self mice. The White Pink-Eye and the White Black-Eye. White mice are judged much more strictly than other colours. For example, a white mouse must be at least thirteen to fourteen centimetres long, both its body and it's tail. The white must be a real snow white.

## Recognised markings and liveries

Mice come in a number of markings and liveries. The markings can each be recognised in certain colourings. The difference between a marking and a livery is that the pattern of a livery is genetically established, in contrast to the pattern of a marking.
The following example makes this clear: A tan livery is always in the same, right place, the point where the belly and topcoat colours meet (the pattern) is established genetically. With a Dutch marking, however, it is very usual that the pattern is disturbed and that the markings are found in the wrong places.

The Tan livery is known from the rabbit world. This livery is made up of two colours. The main colour is on the back and the tan colour on the belly. Recognised main colours for the tan livery are black, blue, dove-grey, chocolate and champagne. The tan colour is a rust-coloured reddish-brown. The darker the main colour, the darker the tan. In any event the tan colour should be as fiery as possible.

The Silver Fox livery is also known from rabbit breeding. There are two colours involved here also, one on the back and the other on the belly. The recognised main colours are black, blue and chocolate. The belly is always

Black and tan

white. Although difficult to breed, white hairs in the coat must be avoided as far as possible. White hairs may appear behind the ears and around the tail.

The Siamese is a livery that is seldom seen but extremely pretty. Like the Russian the body extremities (snout, ears, tail and tail-root) are darker or stronger in colour that the rest of the body. The transition between the colours must be gradual. The Siamese livery is recognised in two colours: brown (seal-point) and blue (blue-point). The Seal-point Siamese is a light sepia colour with brown extremities and ruby-red eyes. The Blue-point Siamese is light blue with medium to dark blue extremities and also with ruby-red eyes.

The Russian, like the Siamese, has extremities of a different colour. But on this type the transition between the colours is much more sharply defined. The mask must cover the whole head, and the tail, ears and paws are also coloured. The main coat is a light white. Recognised livery colours are black, blue and chocolate.

The Dutch is another marking known from rabbit breeding. The head marking consists of two spots around the eyes. The front of the body is white, and the rear of another colour. The border between the fore- and hindquarters is sharply and evenly defined. The tail is also of two colours, with a white tip. Recognised marking colours are agouti, black blue and chocolate.

The Rump-white is a difficult marking to breed. The true-bred form dies at an early age, so it always has to be bred with impure animals to get the marking. On the rump-white the front part of the animal must be coloured and the hindquarters and tail white. The transition is sharply defined. The marking is permitted in all recognised mouse colours.

The Broken Marked mouse is multicoloured. Spots are splashed unevenly across the whole body, must be visible on both sides of the body and may not overlap each other. It must have a nose spot, two differently coloured eyes are permitted.

On the Even Marked mouse, the spots must be splashed regularly.

A newer marking: Variegated

Banded

They must also be sharply defined and not merge into each other.

A special marking is the Banded mouse. This marking comes from cattle breeding. Just like a banded cow, this mouse possesses a dark band around its middle. This band must be approximately three centimetres wide. Like many varieties, this marking originated in the UK. The first banded mice were put up for recognition in the early 90's in the colour chocolate. One year later the colours black and champagne were provisionally approved.

Even now, more and more new hair structures, colours, markings and liveries are appearing, such as curly hair, dapple-grey and tortoiseshell. Since breeders across the world are working daily to "find" new varieties, there'll be a few more surprises yet.

Some of the newer colours and markings: Fawin, Silver, Variegated, Himalyan, Tricolor, Cinnamon, Sable, Marten Sable, Pearl, Silver Grey, Astrex, Seal Point and Brindle.

# Reproduction

**Mice are famous, or rather infamous, for their reproductive powers.**

Of course, it's nice to breed a litter of mice, but be sure in advance that the young have good homes to go to, because they have to leave their mother after about six weeks. Ask your pet shop whether they need young mice. You may also be able to sell or give them to neighbours, friends or acquaintances.

Should you end up having to keep the young, you'll have to find another (perhaps less pleasant) way to deal with them, because you really can't keep them all. The cage will quickly become too small. One solution, of course, is to go out and buy another few cages to house the young, but that's not really an option for many people. So only start to breed if you've found homes for the young.

## In-breeding

To breed responsibly, you must only use strong and healthy adult animals. And you can't put any male together with any female. This would cause a high risk of in-breeding. If you've been given a brother and sister from a neighbour's litter, for instance, it's not advisable to breed them. Mating these animals together is a serious form of in-breeding. And who can guarantee that your neighbour's litter was not also the offspring of a brother and sister? Most rodent varieties live in the wild in relatively isolated groups. Scientists have long questioned whether this way of life does not automatically lead to wide-spread in-breeding. In-breeding occurs when a female mates with a member of its own family (father, bro-

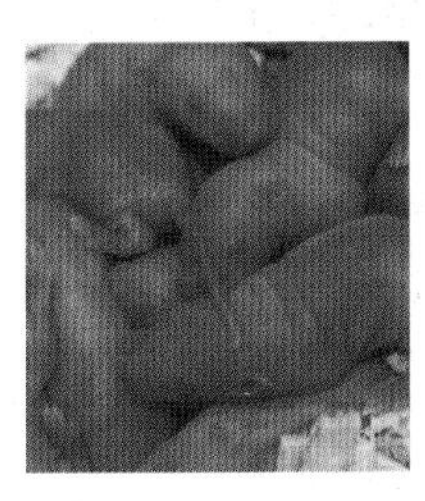

ther, uncle or nephew). In-breeding is harmful because the breeding of animals with the same genes leads to a substantial deterioration in variation. Not only do animals look more and more like each other externally, but they also become more and more one-sided internally. In practice, inbred animals (except for very selective, targeted breeding) become smaller, less fertile and more vulnerable to disease and genetic abnormalities.

Scientific research into mice has shown that in-breeding almost never, if at all, occurs in wild populations. Females in season seem to leave the colony to mate with males from neighbouring colonies. They then return to their own community to bring their offspring into the world and bring them up together with uncles and aunts.

## Male or female

If you want to breed a litter, you first need to be sure you have a male and a female available. The difference between mice of different sexes cannot be seen at a glance. You have to examine them closely under their tail. Like most rodents, you can tell the difference in sexes by the distance between the anus and the genital ope-

Pregnant mice can become very fat

ning. This distance is much wider on males than on females. On adult males you can also detect the circumference of the scrotum.

## Season and mating

Mice are fertile at a very young age. The first signs of sexual maturity appear from about four weeks. On young females, the vagina is closed by a membrane (of flesh) that spontaneously disappears after about 35 days. Then their first ovulation occurs, so mating may be successful from six to ten weeks. If the female remains unmated, she will go into season every four to six days, and is then fertile for twelve to fourteen hours. A male will smell when a female is ready to mate, and will sniff and lick her. The male will run behind the female until she stands still. She then raises her hindquarters and moves her tail to the side. Mating itself takes only a few seconds and is regularly repeated.

## Birth and development

After successful mating, the young are born after a pregnancy of 19 to 21 days. Towards the end of the pregnancy the female's weight can almost double.

A female's first litter usually consists of eight or nine young. The fourth and fifth litters are the largest with ten to twelve young.

These are averages of course. The litter size reduces after approximately six months. Females often become infertile after approximately eighteen months. Males remain fertile for much longer.

After birth, the female goes into season again within 24 hours. She may mate again if the male is not removed, and thus young could be born every three weeks. This is also bad for the mother's health!

Young mice are real home-lovers. In contrast to animals that flee the nest early, they are born naked, blind and helpless. Eyes and ears are shut, and at birth a mouse weighs only 1.5 grams. The first fur appears after three days, and the ears open a day later. The eyes remain closed for twelve to fourteen days. The young drink from their mother for three to four weeks, but at about two weeks they will start to nibble at solid food.
If you leave the young in a group, the adult males will start to mate with the young females from the most recent litter after about six to ten weeks. This is not only bad for the young females, but is also in-breeding. You will then be saddled with more young, so you must take the young from the group after five to six weeks, or separate the males from the family.

# Other species

**The fancy mouse has been known and loved as a pet for many years. But there are many other species of mice, which are more or less related to the fancy mouse and also very suitable as pets.**

Jerboa's are jumping mice

Because there are far fewer examples of these varieties in captivity, the populations have to be handled very carefully. Co-operation between owners of other species can ensure that these strains are gradually and responsibly expanded, so that more enthusiasts can enjoy these attractive animals.

Just like the fancy mouse, most special mouse varieties enjoy life in captivity if they are kept in a responsible manner. This chapter will introduce you to a number of special mouse varieties kept by enthusiasts.

## The Harvest Mouse

Close to home, we find the Harvest Mouse *(Micromys minutus)* in the wild. This mouse is found in a huge area stretching from the British Isles to deep into Russia and Turkey. The harvest mouse is not found in Scandinavia, Spain, Southern Italy, Greece or Southern Turkey. This mini-mouse is Europe's smallest rodent. It is only five to seven centimetres long and weighs between four and ten grams. The harvest mouse is somewhat plump in form and its body has totally evolved to become a climber. Its tail is approximately the same length as its body.

Harvest mice prefer life in areas with corn or reed fields, shrubs, meadows or hedges. They spend just about their whole life climbing there. Only in the cold winter months, when there are no grasses or areas to climb in, do they seek out holes or barns to spend the winter in. Harvest mice do not hibernate as such, but they are much less active in the winter months.

During the summer months, harvest mice are practically never found on the ground. They balance from one stem of grass to the next. It's fascinating to watch this little animal climbing. It spreads its toes out wide and closes its grip on the stem, using its tail as a support and brake. The busy harvest mouse is active both during the day and for short periods at night.

The reproductive behaviour of the Harvest Mouse is especially interesting. Harvest mice build a pretty, ball-shaped nest in grass. A group of mice can build such a nest in a couple of days. It is some eight centimetres wide and anchored to a pair of grass or corn stems, from twenty centimetres up to one metre above the ground. The mice gather standing stems together and then fix them firmly together. Then they weave the nest from freshly bitten-off stems and leaves. The walls are coated on the inside with very soft, finely chewed plant fibres. In good conditions, a female may bear six litters, but usually it is less. After a pregnancy of about three weeks, four to six young are born, sometimes as many as nine. The young are nest-lovers. They are bald,

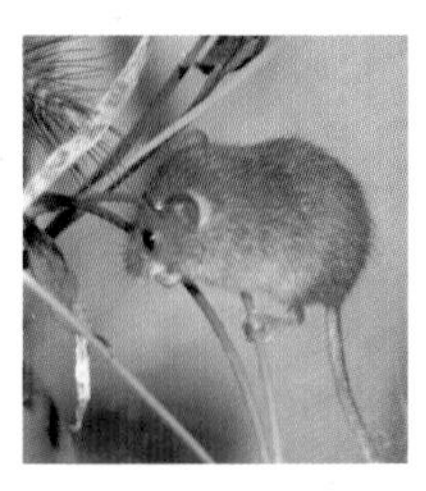

Harvest Mouse

Harvest Mouse

Harvest Mouse

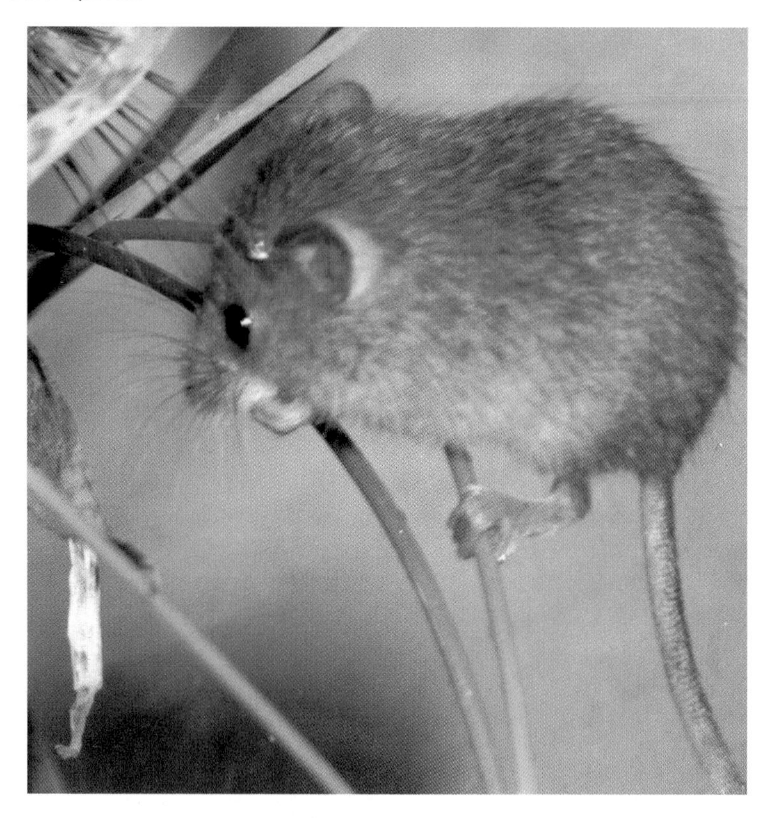

blind and helpless, two centimetres long and weigh less than one gram. Their eyes open after eight to ten days. Very quickly, the young are clambering among the stems around the nest. They can now eat on their own, but normally are still suckled by their mother for a few days more. They spend their youth climbing merrily and perilously and playing other games. The young are sometimes sexually mature at five to six weeks. Harvest mice can reproduce rapidly. So much the better, because in the wild they are very vulnerable. Large numbers fall prey to agricultural machinery, predators and the cold winter.

In the wild, the harvest mouse feeds primarily on grass and herb seeds, grain and buds. Sometimes they also eat insects and insect larva.
The Harvest Mouse is simple to keep and nice, active pets.
They're not rare and have absolutely no problem with life in captivity but, like in the wild, they are fragile. If you give a group of harvest mice a nice, nature-like home

with plenty of opportunities to climb, they will be grateful pets. The examples on sale are generally not caught in the wild, but bred by enthusiasts. This population of harvest mice has become relatively domesticated. It's unwise to capture harvest mice in the wild and then keep them as pets.

### The Striped Grass Mouse

The Striped Grass Mouse *(Lemniscomys barbarus)*, also known as the Barbary Striped Mouse or Zebra mouse, is kept with mixed success as a pet. Some breeders have been successful, others never have a litter. Striped Grass mice are pretty, active animals, and make excellent pets. But they're not animals to cuddle. As long as you leave them in peace and just observe them, everything is fine, but they are shy and too fast to pick up. The Striped Grass mouse is active during the day, but especially in the dawn or dusk twilight.

This mouse's name says a lot about its appearance and its habitat. It is indeed striped and lives mainly in grasslands. In central and southern Africa, where this mouse lives, the grass is not green, but brown. The lighter and darker stripes on its back make the Zebra mouse practically invisible between stems of grass. It is almost as big as the house mouse and of the same build. Its body is nine to fourteen centimetres long, and its tail ten to fifteen centimetres. Grass mice are extremely fast and agile. If you open a cage of Zebra grass mice, the animals will fly off in all directions. This is a natural escape reaction.

Striped Grass Mouse

Striped Grass Mouse

Striped Grass mice are best kept in groups, but with not too many adult males. They need a lot of space and a cage of 100x40 cm is the minimum size. The cage must be heated in the cold winter months, and the temperature must never drop below 12° C. A lightbulb over part of the lid is a good idea. You should fit out the cage with lots of hideaway opportunities and large bundles of hay. A stone to sit on, or block of wood near the window is also advisable. Grass mice benefit from a varied diet. Grasshoppers form an important part of this. Not only the animal protein, but also the hunt for these insects seems to play an important role in getting the mating instinct working. At twilight the male starts chasing the female and will keep up the chase until the female offers herself. Mating is a furious activity, where both animals fall around, and is repeated a number of times. After a pregnancy of some 28 days, two to five young are born that develop quickly. Their eyes open after just one week. You must never touch the young, as Striped Grass mice tend to eat their young if they detect a foreign scent.

### The African Spiny Mouse

The African Spiny Mouse has been kept as a pet in Europe for some ten years, particularly in Germany. They come in various varieties and sub-varieties. In general, they are mice with a normal to strong build and a typical spiny coat. If you stroke its back from back to front, you can clearly see and feel the spines. Actually, this mouse is a small hedgehog with a tail. Spiny mice are nice, interesting animals, which can generally be easily kept as pets. They don't have special needs in terms of housing and show themselves during the day as well. They're best kept in groups. Such a group may be a pair together with two young females that are not related. If you change the make-up of a larger group after a while, you must expect fights for supremacy, which may cost a couple of mice their lives. Almost all varieties and sub-varieties of the Spiny Mouse live in North Africa and the Middle East.

Reproduction of the various varieties is practically identical. After a pregnancy of 36 to 38 days, the young are born, usually in the morning hours. A litter usually consists of one to three young. After this relatively long pregnancy the young are much further developed than those of many other rodents. They're quite big (five to seven grams), and their eyes are almost or completely open. They will scramble around the cage within a day or so. Spiny mice do not make a nest. Birth and care for the young is a group matter. All the females adopt the role of a wet nurse. If there are several females with milk available they will also help suckle the young. The suckling period is over after five or six days. In the wild, spiny mice never live longer than eighteen months to two years, but in captivity can live up to more than three years. African spiny mice are not fussy as far as their diet is concerned. They eat seeds, grain, bread, vegetables and fruit, but also enjoy the odd mealworm. They can live without water, but do like it if available.

**The Sinai Spiny Mouse**

The Sinai Spiny Mouse *(Acomys dimidiatus)* is kept in varying quantities as a pet. Sometimes a relatively high number are kept and they breed well, sometimes smaller numbers are kept. This spiny mouse is mouse-like in form with a pointed head and prominent eyes. The topcoat is brownish with ticking. The belly is a sharply defined white. Ears are very dark. Sinai spiny mice are found in Israel and Egypt.

Yellow African Spiny Mouse

African Spiny Mouse

Black Nile Spiny Mouse

**The Golden Spiny Mouse**

The Golden Spiny Mouse *(Acomys russatus)* is easily recognised by its posture and format. Because its body is so plump it also behaves differently to a real mouse. It is golden brown in colour. Its eyes are fairly small and its paws, toes and ears are covered with fur. The black soles of its feet are striking. It can put its spines up as a defensive weapon. The golden spiny mouse lives on the rocky slopes and deserts of the Middle East. It's a good rock-climber, and can usually be seen in the twilight of dawn or dusk, as it avoids the highest temperatures during the day. The golden spiny mouse can jump, but not very far or high. It's a peace-loving animal.

Mating is also a fairly peaceful affair. The male will chase the female until she stands still. He uses one paw to move the females tail to the side and then mates with her.

**The Egyptian and Black Nile Spiny Mouse**

Both the yellow Egyptian Spiny Mouse *(Acomys cahirinus)* and the Black Nile Spiny Mouse *(Acomys cahirinus cahirinus)* are regularly kept as pets in Western

Europe. The original form of the strain is yellow to golden in colour and has big, round eyes. The spines on its back are not really hard and are unsuitable as a defence weapon. It has a pink snout and its paws and toes are thinly haired. Its fur is especially glossy. The animal is mostly found in Egypt. Just like the house mouse, it is a culture follower and lives in houses and temples. Egyptian spiny mice behave in many ways like normal mice. They are extremely fast and agile and can jump very high from standing. They sometimes live in shrubs, where they race along the thinnest branches like tightrope artists. They sometimes also sleep on thin branches. With even the slightest grip available, they can run up walls. Egyptian spiny mice are only active at nights. Their escape behaviour is striking. If they're approached from above, such as by a predatory bird, they shoot into their holes. If approached from the ground, as a snake would do, they spring around all over the place. When mating the male chases the female at unbelievable speeds. When the female stops the male will bite her at the tail root, so that she lifts her hindquarters to allow mating.
The Black Nile Spiny Mouse is a so-called melanotic sub-specie of the Egyptian spiny mouse. Melanotism is a black mutation of an originally light-coloured animal. The topcoat colour is almost black, although that is not easy to see in the light because of their high gloss. On this black sub-specie the belly is sometimes dark, sometimes white and sometimes in-between

### The Japanese Dancing Mouse

Now and then, you may come across a mouse called the Japanese Dancing Mouse. This animal is characterised by its comical movements. It looks just like the mouse is dancing. However, Japanese Dancing Mice are nothing other than fancy mice with an inherited brain abnormality. These animals can no longer move normally. Breeding of such animals should be forbidden; it is a form of cruelty.

### The Hairless Mouse

The Hairless Mouse is another sad example of irresponsible breeding to create special animals. A number of laboratories have succeeded in breeding fancy mice with almost no hair. They're as good as naked. These animals look like new-born mice, but larger. They are a shrivelled pink with extremely scarce thin white hairs. These mice may be valuable for scientific research but surely don't belong kept at home as a pet or a curiosity!

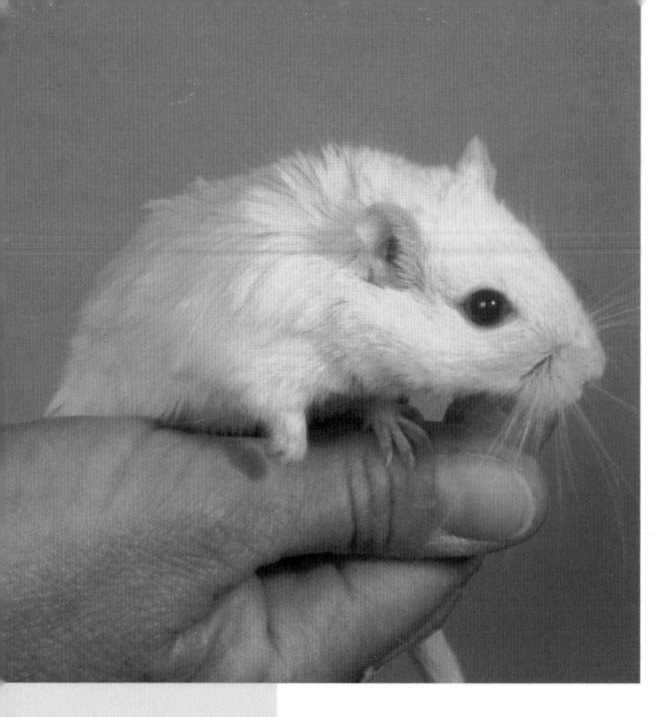

# Your mouse's health

**Fortunately, mice generally have few health problems. A healthy example has bright eyes and is lively.**

Its coat is smooth, soft and regular. Its rear body is dry and clean. A sick mouse sits withdrawn all the time. Its coat is dull and stands open, as if wet. The animal's back is raised, even when walking.

## Prevention

The rule that "prevention is better than a cure" also applies to small animals such as the mouse. It's not always easy to cure a sick mouse. They are so small that even a vet doesn't always know how to treat them.

Even a light cold can prove fatal for a mouse and the biggest risks to its health are draughts and damp.

There are a few general rules that you can follow if your mouse is ill:

- Keep the animal is a quiet semi-dark place. Stress, crowding and noise won't help it get better.
- Keep the animal warm, but make sure its surroundings are not too hot. The best temperature is 20 to 21ºC.
- Don't wait too long before visiting a vet. Small rodents that get sick usually die within a few days.
- The patient should always have fresh water and remember that your animal may be too weak to reach its water bottle.
- Sick animals often eat little or nothing. Give it a small piece of apple or other fruit.

## Colds and pneumonia

Draughts are the most common cause of colds and pneumonia for mice, so choose the place for its home carefully.

They can withstand low temperatures, but cold in combination with a draught almost inevitably leads to a cold. The mouse starts sneezing and gets a wet nose. If its cold gets worse, the animal starts to breathe with a rattling sound and its nose will run even more, so it's now high time to visit the vet, who can prescribe antibiotics. A mouse with a cold or pneumonia must be kept in a draught-free and warm room (22 to 25° C).

### Diarrhoea

Diarrhoea is another formidable threat to mice and often ends fatally. Unfortunately, diarrhoea is usually the result of incorrect feeding, sometimes in combination with draughts or damp. Most cases of diarrhoea are caused by giving the animal food that is too moist. Rotted food or dirty drinking water can also be a cause. You can do a lot to prevent diarrhoea.

But should your mouse become a victim, then take any moist food out of the cage immediately. Feed your animal only dry bread, boiled rice or crispbread. Replace its water with lukewarm camomile tea. Clean out its cage litter and nest material twice a day. As soon as the patient is completely recovered, you must disinfect its cage.

### Wet tail

E-coli bacteria cause an especially serious form of diarrhoea and most victims die within 48 hours.

Mice that fall victim to this disease have a constantly wet tail and anus; they won't eat and become apathetic.

E-coli bacteria are normally present in small quantities in the intestines of a small rodent. In the event of reduced resistance or stress, the bacteria suddenly become active. Whenever your mouse has a wet tail take it to the vet's immediately.

## Tumours

Mice relatively often have problems with tumours. Tumours occur more frequently in strains where in-breeding has occurred, in other words where animals have been crossed with their own family members. The most common tumours affect the female's teats, but tumours can also be the result of skin cancer. These forms can be operated on but, because of the animal's age, this rarely makes sense.

Another form of tumour is caused by infections under the skin, which is called an abscess. A small wound may heal, but an infection remains under the skin. This type of tumour can be easily treated by a vet who opens and cleans it. Should your mouse show signs of a tumour, take it straight to the vet's. Delaying can only make things worse, both with skin cancer and abscesses.

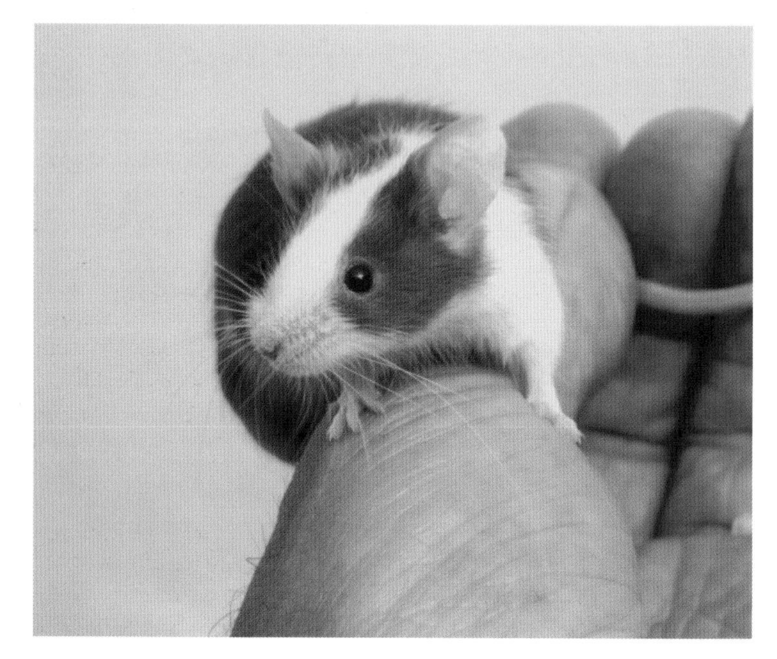

## Broken bones

Mice sometimes break bones because they get stuck with their paws, jump off your hand or fall from a table. An animal with a broken paw will not put weight on it and will limp around the cage.

If it's a "straight" fracture (the paw is not deformed), this will heal within a few weeks. Take care that the mouse can reach its food and drink without difficulty. If a mouse has broken its back, it's best to have it put to sleep. If in doubt about a possible fracture, always ask your vet.

## Overgrown teeth

A rodent's front teeth grow continuously and are ground regularly by its gnawing. A genetic defect, a heavy blow or lack of gnawing opportunities can disrupt this process. Its teeth are ground irregularly and in the end don't fit together properly. In some cases the teeth continue to grow unchecked, even into the opposite jaw.

When a rodent's teeth are too long, it can no longer chew properly and the animal will lose weight and eventually starve to death.

| Deficiency of | Symptoms | Found in |
| --- | --- | --- |
| **Protein** | Poor coat, hair loss, pneumonia, infertility and poor growth of young animals, aggression (both with too much and too little) | Peas, beans, Soya, cheese |
| **Vitamin A** | Pneumonia, damage to mucous membrane or eyes, growth problems, diarrhoea and general infections, cramps, small litters | Root vegetables, egg-yolk, fresh greens, bananas and other fruit, cheese |
| **Vitamin B complex** | Hair loss, reduced fertility, weight loss, trembling, nervous symptoms, anaemia, infections | Oat flakes, greens, fruit, clover, dog biscuits, grains |
| **Vitamin C** | The mouse produces this itself, deficiency rarely a problem Growth problems, poor bone condition | Greens, fruit |
| **Vitamin D** | Too much vitamin D causes calcium loss in bones and calcium deposits in blood vessels | Dairy products, egg-yolk |
| **Vitamin E** | Infertility, muscle infections, nervous problems, bleeding and poor growth of young animals | Egg-yolk, sprouting grains, fresh grains, greens |
| **Vitamin K** | (Nose) bleeding, poor healing of wounds and growth problems. | Greens |
| **Calcium** | Normally produced in the animal's intestines. Lameness, calcium loss in bones and broken teeth | Mineral preparations, dairy products, sepia, varied diet |
| **Potassium** | Weight loss, heart problems and ascitis , wetness in open abdominal cavity | Fruit |
| **Sodium** | Can only occur with serious diarrhoea | Cheese, varied diet |
| **Magnesium** | Restlessness, irritability, cramps, diarrhoea and hair loss | Greens, grains |
| **Iron** | Anaemia, stomach and intestinal disorders, infertility | Greens, grains, meat |
| **Iodine** | Metabolic disorders and thyroid gland abnormalities | Greens, grains, water |

Long teeth can easily be clipped back. A vet can show you how to do that, or do it for you if you don't feel able. Take care that your rodent always has enough to gnaw on. A piece of breeze block, a block of wood or a branch will do fine.

## Malnutrition ailments

Not only calcium deficiency, but also a shortage of other minerals and vitamins can lead to sicknesses. See the table on page 56 for an overview of sicknesses that can arise from certain deficiencies.

## Parasites

Parasites are small organisms that live at the cost of their host. The best known are fleas on dogs and cats. Rodents seldom have problems with parasites, and certainly not healthy animals. Weak, sick or poorly cared for animals, however, are far more likely to be affected. You mostly discover parasites only when your animals start to scratch

themselves and get bald patches. If you notice that your mice are itching and scratch themselves frequently, then they're probably suffering from lice (tiny spiders that feed on blood). These lice are often spread by birds, and mice sometimes pick up a flea from a dog or cat.

Your pet shop or vet can advise you on dealing with parasites.

### Skin mites

The skin mite is a particularly harmful parasite. Fortunately they seldom occur but if they affect your mouse colony, you've got work to do. The skin mite is a minute spider that creeps into its host's skin, making the mite itself almost never visible. It causes scabs and eczema, which sometimes cover the whole skin within a month. Skin mites are infectious and can be passed on to other animals.
Your vet or a good pet shop will have treatments for skin mites. Read the instructions on the packaging thoroughly. In most cases the infected animal must be bathed in the substance. Dry off the mouse well to prevent it catching a cold and put it in a warm place (minimum 25° C).

### Fungal skin infections

Rodents can sometimes suffer from fungal skin infections, which cause small areas of flaking in the ears or nose. Skin fungi are infectious to humans and animals but easy to treat. But don't let the problem go on too long because the animal may suffer other ailments because of it. Your vet has standard medicines against fungi.

### Old age

Obviously we hope that your mice will grow old without disease and pain. However, mice get nowhere near as old as humans and you must reckon with the fact that after two years you have an old mouse to care for. Such an old mouse will slowly become quieter and get grey hair in its coat or lose hair. Such an old animal needs a different kind of care. The time for wild games is over; it won't like them any more. Leave your mouse in peace. In the last few weeks and days of its life, you will notice its fur decaying and the animal will get thinner. Don't try to force it to eat if it doesn't want to; the end is usually not long off. Mice, on average, live only three years. A four-year old mouse is very old.

# Tips for the mouse

- Take the time to visit a small animal show.
- Never set a pet rodent free. It will not survive!
- Mice love to climb and clamber. Fit their cage out with good climbing opportunities!
- Mice must always have fresh water available
- Some special mouse types also make good pets.
- Mice are herd animals. They will be really unhappy sitting alone in a cage.
- Take rotting food out of the cage. It can cause disease.
- Draughts and damp are the biggest threat to your rodent's health.
- Never buy a mouse that has been taken from its mother too early.
- Fancy mice can breed rapidly. Think about that when putting your group together.
- Avoid in-breeding, not just with mice, but with all animals!
- Fancy mice are not nasty or creepy!
- Never give your pet too much vegetable. Salad and cabbage can upset their intestines.
- Never buy a pet at an animal market.

# Clubs

**Becoming a member of a club can be very useful for good advice and interesting activities. Contact the a Club in case addresses or telephonenumbers are changed.**

**National Mouse Club (UK)**
The NMC is purely a fanciers' organisation, dedicated to those who breed mice for showing. It is not aimed at pet keepers.

Sec. Brian Cookson
44 Speeton Avenue
Bradford
West Yorkshire BD7 4NQ
Email:annandbrian@talk21.com
Tel: 01274 574205

**London & Southern Counties Mouse & Rat Club**
Holds mouse & rat shows on the first Saturday of every month at the 4th Enfield Scout Hall, Gordon Rd, Enfield.
For details, contact the Hon.

Sec.Alan Birch
426 Goresbrook Road
Dagenham
Essex RM9 4UX.
Email: alan@birch9972.fsnet.co.uk

# The Mouse

| | |
|---|---|
| Latin name: | *Mus musculus, Linnaeus* 1758 |
| Family: | Muridae |
| Origin: | Originally a narrow area between Spain and Asia. Spread throughout the world by man. |

**Measurements and weight**

| | |
|---|---|
| Body length: | 69-100 mm |
| Tail length: | 58-90 mm |
| Weight: | 10-33 grams |

**Reproduction**

| | |
|---|---|
| Sexual maturity: | 6 weeks |
| Pregnancy: | 19-23 days |
| Number of young: | 4-8 |
| Eyes open: | 12-15 days |
| Suckling period: | 4 weeks |
| Life expectancy: | 3, max. 20 months (in the wild)<br>3, max. 4 years (as a pet) |